PYTHON PROGRAMMING

The Fundamental Beginner's Guide to Learning Python

THOMAS JACKSON

Copyright

All rights reserved. No part of this publication may be reproduced, stored in a retrieval system or transmitted in any form or by any means, electronic, mechanical, photocopying, recording and scanning without permission in writing by the author.

Printed in the United States of America

©2019 by Thomas Jackson

About the Author

Thomas Jackson is a tech enthusiast with about 10 years' experience in the ICT industry. He is passionate about the latest technical and technological trends. Thomas holds a Bachelor and a Master's Degree in Computer Science and Information Communication Technology respectively from MIT, Boston Massachusetts.

Who Is This Book Aimed At?

This book is best suited for beginners. It is intended for anyone who so far has not engaged seriously in programming and would like to begin doing it. This book starts from scratch and introduces you step by step into the fundamentals of programming. It won't teach you absolutely everything you might need for becoming a software engineer and working at a software company, but it will lay the groundwork on which you can build up technological knowledge and skills, and through them you will be able to turn programming into your profession.

If you've never written a computer program, don't worry. There is always a first time. In this book we will teach you how to program from scratch. We do not expect any previous knowledge or abilities. All you need is some basic computer literacy and a desire to take up programming. The rest you will learn from the book.

Table of Content

Copyright --- iii

About the Author --- iv

Who Is This Book Aimed At? ------------------------------------ v

INTRODUCTION TO PYTHON ------------------------------------ 1

   1.1 Why Choose Python? -- 2

GETTING STARTED --- 3

   2.1 Downloading Python --- 4

   2.2 Choosing a Text Editor ------------------------------------- 5

   2.3 Downloading PyCharm -------------------------------------- 5

   2.4 Running your First Program -------------------------------- 7

VARIABLES AND DATA TYPES -------------------------------- 12

   3.1 What is a Variable? -- 12

      3.1.1 Creating a Variable ------------------------------------ 13

3.2 Python Data Types -- 15

    3.2.1 Strings -- 16

    3.2.2 Numbers --- 17

    3.2.3 Boolean -- 18

    3.2.4 List -- 18

    3.2.5 Dictionary --- 19

    3.2.6 Tuples -- 21

OPERATORS -- 23

  4.1 What Is an Operator? --------------------------------- 23

  4.2 Arithmetic Operators ---------------------------------- 24

  4.3 Logical Operators -------------------------------------- 25

  4.4 Comparison Operators -------------------------------- 26

  4.5 Assignment Operators --------------------------------- 27

  4.6 Identity Operator --------------------------------------- 28

  4.7 Membership Operators ------------------------------- 29

  4.8 Bitwise operators --------------------------------------- 30

Exercises -- 31

CONDITIONAL STATEMENT ---------------------------------- 32

    5.1 The "if" statement -- 32

    5.2 The "If-Else" statement ------------------------------------- 34

    5.3 The "If-Elif-Else" Statement ------------------------------ 36

    Exercises -- 38

LOOPS -- 39

    6.1 Types of loops in Python --------------------------------- 41

        6.1.1 While Loop --- 41

        6.1.2 For Loop --- 46

        6.1.3 Nested Loops -- 49

    Exercises -- 52

ARRAYS -- 53

    7.1 Creating Arrays -- 53

    7.2 Access to the Elements of an Array ----------------- 55

    7.3 Basic Array Operation ------------------------------------- 55

    7.3.1 Length of An Array --------------------------------------- 56

    7.3.2 Add Elements to Array ---------------------------------- 56

    7.3.3 Remove Elements from Array ------------------------- 57

    7.3.4 Array Concatenation ------------------------------------- 59

    7.3.5 Slicing Array -- 59

    7.3.6 Iterating through an Array ------------------------------ 60

FUNCTIONS -- 63

  8.1 Importance of Function -------------------------------------- 63

  8.2 Function Declaration -- 64

  8.3 Parameters in Functions -------------------------------------- 65

    8.3.1 Functions with Multiple Parameters ----------------- 67

  8.4 Returning a Result from a Function ---------------------- 68

  8.5 Default Arguments -- 70

  8.6 Anonymous Functions --------------------------------------- 71

    8.6.1 How to write Anonymous Functions ---------------- 72

    8.6.2 Lambda within user-defined functions ------------- 72

8.7 Scope of Variables --- 73

FILE HANDLING --- 75

9.1 How to open a file? --- 76

9.2 Reading File --- 77

9.3 Creating a New File --- 79

9.4 Delete a File --- 80

OBJECT-ORIENTED PROGRAMMING (OOP) --- 81

10.1 Fundamental Principles of OOP --- 81

10.2 Classes --- 83

10.2.1 Attributes and Methods in class --- 84

PYTHON CONSTRUCTOR --- 87

11.1 Creating the constructor in python --- 87

11.2 Python In-built class functions --- 90

11.3 Built-in class attributes --- 91

INHERITANCE --- 94

Sub-classing --- 96

Constructors with Inheritance ----------------------------------- 96

CONCLUSION -- 97

INTRODUCTION TO PYTHON

Python is a loosely typed high level programming language which can be used to perform to perform programming tasks ranging from web and desktop application development to data science and machine learning. Unlike many similar languages, its core language is very small and easy to master, while allowing the addition of modules to perform a virtually limitless variety of tasks.

Python was created by Guido van Rossum in late 1980s at the National Research Institute for Mathematics and Computer Science in the Netherlands. The initial version was published in 1991, and version 1.0 was released in 1994. Python is a true object-oriented language, and is available on a wide variety of platforms.

This book provides a gateway to in-depth Python programming.

1.1 Why Choose Python?

There are several features that make Python an appealing choice. Some of them are as follows:

- Simplicity:
- Portability:
- Open Source and Large Support
- Used for Data Science and Machine Learning:
- Support for Web development:

GETTING STARTED

In this chapter you will learn how to download and install Python on your computer, run its built-in programming environment, and create your very first Python program. You can simply install core Python and use a text editor like notepad to write Python programs. These programs can then be run via command line utilities. The other option is to install an Integrated Develop Environment (IDE) for Python. IDE provides a complete programming environment including Python installation, Editors and debugging tools. Most of the advanced programmers take the IDE route for Python development. We are also going to take the same route.

Pycharm is the IDE that we are going to make use of throughout this book.

2.1 Downloading Python

The easiest way to install Python is to download an installer for your computer from the official Python website, https://www.python.org/downloads.

Python 3.8 is the latest version, as of December 2019. The coding exercises in this book assume that you are using this version of python 3, or at least a version close to it.

Installing Python is easy and straightforward, and should proceed without any problem.

2.2 Choosing a Text Editor

The next step is to choose a text editor, since you are going to be writing a bunch of codes, you are going to need some environment of some program that you can write all that code in. You can write python in any text editor like a notepad, it doesn't really matter but there are special text editors that are designed just for writing python codes, and these are called IDE (Integrated Development Environment).

It's basically just a special environment where we can you run and execute our python code. It basically like tell us how we're doing so were the only if we write something that's wrong or we have errors it'll kind of point us in the right direction of what we need to do to fix it.

In this book we are going to make use of PyCharm.

2.3 Downloading PyCharm

To install PyCharm, head over to this website https://www.jetbrains.com/pycharm.

When you click on the **DOWNLOAD** button, you will get this page that shows two versions of PyCharm.

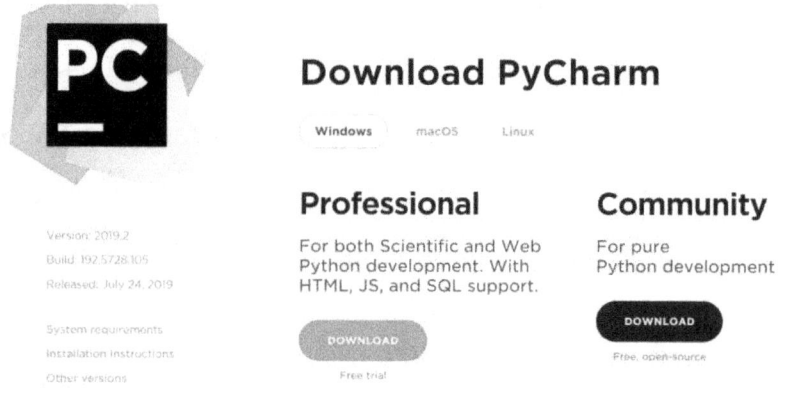

One is a professional version which is a paid version. The other is the community version and it is free and open source. Download the community version, it has everything we need to get started using python.

Installing PyCharm is also easy and straightforward, and should proceed without any problem.

2.4 Running your First Program

We have installed environment required to run Python scripts. Now we are going to run our first python program to see how it works.

First thing is to open PyCharm editor. You can see the introductory screen for PyCharm. To create a new project, click on "Create New Project".

You will need to select a location.

1. You can select the location where you want the project to be created. If you don't want to change location than keep it as it is but at least change the name from "untitled" to something more meaningful, like "MyProject".
2. PyCharm should have found the Python interpreter you installed earlier.
3. Next Click the "Create" Button.

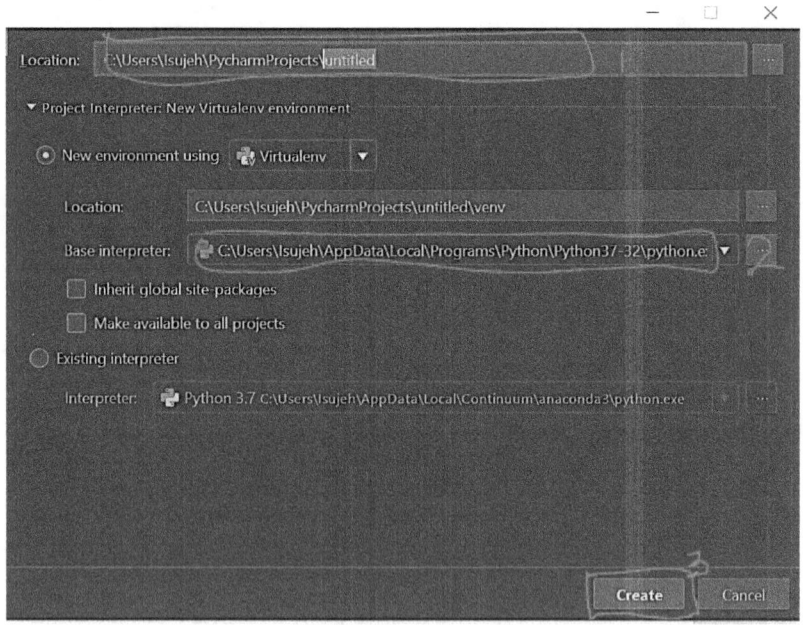

Now under the project folder that you have just created, right click select "New" then, select "Python File".

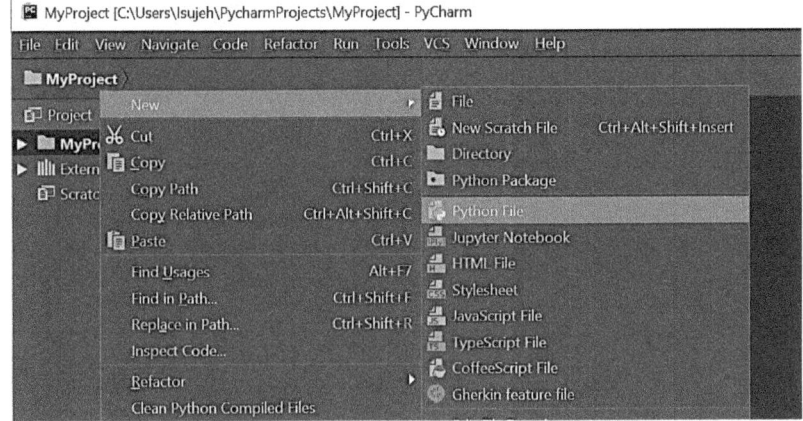

A new pop up will appear. Now type the name of the file you want (Here we used "HelloWorld") and hit "OK".

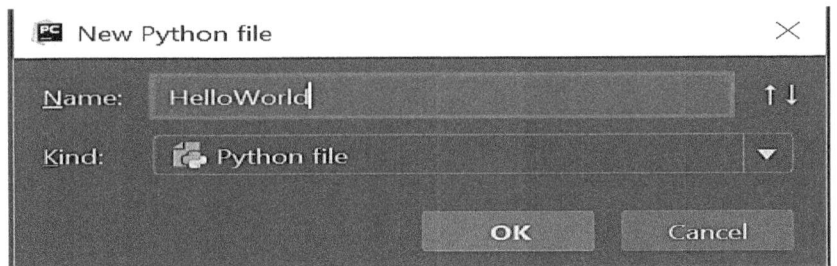

Now type a simple program - print ('Hello World!').

Now Go up to the "Run" menu and select "Run" to run your program.

You can see the output of your program at the bottom of the screen.

That's it for your first program! It's your very first Python program, so tap yourself on the back and congratulate yourself on this achievement.

What's next?

In this chapter we saw the process of setting up the environment required to run python programs. We wrote our first python program using the PyCharm editor.

VARIABLES AND DATA TYPES

This chapter will teach you about variables in Python programming. Here you will learn about variable assignments and the different variable types, including numbers and strings.

3.1 What is a Variable?

Variable in programming is a reserved memory location used to store some value. Whenever you store a value in a variable, that value is actually being stored at physical location in memory. Variables can be thought of as reference to physical memory location. The size of the memory reserved for a variable depends upon the type of value stored in the variable. In python, variables are dynamically typed unlike other languages in which variables are statically typed. This means that in python, when a variable is declared, it can be changed and the variable holds a new value.

3.1.1 Creating a Variable

It is easy to create a variable in python. The assignment operator "=" is used when creating a variable. The value on the left-hand side of the assignment operator is the variable name. The value on the right-hand side of the operator is the value assigned to the variable.

Let see examples.

```
Subject = 'Chemistry'    # A string variable
Age = 15                 # An integer variable
Score = 102.5            # A floating type variable
Pass = False             # A Boolean Variable
```

In the examples above, you notice that we didn't specify the variable types with the variable name. For example, we did not write "string Name" or "int Age". We only wrote the variable name. This is because Python is a loosely typed language.

Depending upon the value being stored in a variable, Python assigns type to the variable at runtime. For instance, when Python interpreter interprets the line "Age = 15", it checks the type of the value which is integer in this case. Hence, Python understands that Age is an integer type variable. To check type of a variable, pass the variable name to "type" function as shown below:

```
Age = 15           # An integer variable
Score = 102.5      # A floating type variable
Pass = False       # A Boolean Variable
print(type(Age))
```

You will see that the above script, when run, prints **<class 'int'>** as the output.

Python allows multiple assignment which means that you can assign one value to multiple variables at the same time. Take a look at the following script:

```
Score = Point = Calc = 20   #Multiple Assignment
print(Score)
print(Point)
print(Calc)
```

In the code above, integer 20 is assigned to three variables: Score, Point and Calc. If you print the value of these three variables, you will see 20 thrice in the output.

3.2 Python Data Types

A programming application needs to store variety of data. Consider scenario of a banking application that needs to store customer information. For instance, a person's name and mobile number; whether he is a defaulter or not; collection of items that he/she has loaned and so on. To store such variety of information, different data types are required. While you can create custom data types in the form of classes, Python provides six standard data types out of the box. They are:

- Strings
- Numbers
- Booleans
- Lists
- Tuples
- Dictionaries

3.2.1 Strings

In python, strings are treated as sequence of characters. Strings in Python are created using either single or double quotes. We use the '+' symbol to join strings together, this is known as concatenation.

Example

```
first_name = 'LeBron' #string with single quotation
last_name = "James" #string with double quotation
full_name = first_name + " " + last_name #string concatenation
print(full_name)
```

In the above script we created three string variables: first_name, last_name and full_name. String with single quotes is used to initialize the variable "first_name" while string with double quotes initializes the variable "last_name". The variable full_name contains the concatenation of the first_name and last_name variables. Running the above script returns following output:

LeBron James

3.2.2 Numbers

The ease with which Python performs numerical calculations is touted by many, as among the language's core strengths. Thus, it is important to understand how numerical variables work in Python. There are four types of numeric data in Python:

- Integers or simply, int (stores integer values e.g. 10)
- float (Stores floating point numbers e.g. 5.9)
- long (stores long integer such as 45678976678)
- complex (stores complex numbers such as 7j+4868k)

To create a numeric Python variable, simply assign a number to variable. In the following script we create four different types of numeric objects and print them on the console. We do not make use of quotation marks when storing numbers.

```
int_num = 56                        # integer
float_num = 3.142                   #float
long_num = 5977485613454646         #long
complex_num = -.75+7J               #complex
print(int_num)
print(float_num)
print(long_num)
print(complex_num)
```

The output of the above script will be as follows:

56
3.142
5977485613454646
(-0.75+7j)

3.2.3 Boolean

Boolean variables are used to store Boolean values. Boolean value can either be True or False. Example below:

```
is_male = True
has_muscles = False
```

In the script above we created two Boolean variables "defaulter" and "has_car" with values True and False respectively. We then print the result of the AND operation on both of these variables. Since the AND operation between True and False returns false, you will see false in the output. We will study more about the logical operators in the next chapter.

3.2.4 List

In Python, the data type List which is ordered and changeable is used to store collection of values. Lists are similar to arrays in other programming languages with a difference, Python lists can store values of different types. That is both numeric and string data can be stored in a python list.

Opening and closing square brackets are used to create a list. The items in the list are separated by comma.

Take a look at the example below.

```
1. # empty list
2. list1 = []
3.
4. # list of integers
5. list2 = [1, 2, 3]
6.
7. # list with mixed datatypes
8. list3 = [1, "Red", 3.4]
```

3.2.5 Dictionary

In python, dictionary is an unordered collection of items. While other compound data types have only value as an element, a dictionary has a key: value pair. Dictionaries are optimized to retrieve values when the key is known.

Opening and closing curly braces {} are used to create a dictionary. The items in the dictionary are separated by comma.

An item has a key and the corresponding value expressed as a pair, key: value. While values can be of any data type and can repeat, keys must be of immutable type and must be unique.

```
1. # empty dictionary
2. dict1 = {}
3.
4. # dictionary with integer keys
5. dict2 = {1: 'Samsung', 2: 'basket'}
6.
7. # dictionary with mixed keys
8. dict3 = {'name': 'Jack', 1: [2, 4, 3]}
```

We can also create a dictionary using the built-in function dict().

```
1. # using dict()
2. a_dict = dict({1:'apple', 2:'ball'})
3.
4. # from sequence having each item as a pair
5. b_dict = dict([(1,'apple'), (2,'ball')])
```

3.2.6 Tuples

A tuple is similar to a python list. The difference between here is that we cannot change the items of a tuple once it is assigned whereas, in a list, items can be changed. A tuple is created by placing all the items inside parentheses (), the items in the tuple are separated by comma. The parentheses are optional; however, it is a good practice to use them. A tuple can have any number of items and they may be of different types (integer, float, list, string, etc.).

```
1.   # Empty tuple
2.   tuple1 = ()
3.
4.   # Tuple having integers
5.   tuple2 = (3, 5, 7)
6.
7.   # tuple with mixed datatypes
8.   tuple3 = (1, "Sample", 3.4)
9.
10.  # nested tuple
11.  tuple4 = ("rat", [8, 4, 6], (1, 2, 3))
```

A tuple can also be created without using parentheses. This is known as tuple packing.

```
1.   new_tuple = 3, 4.6, "mouse"
```

```
2.
3.    # tuple unpacking is also possible
4.    a, b, c = first_tuple
```

OPERATORS

All programming language uses operators, through which different actions can be performed on the data. Let's take a look at the operators in Python and see what they are for and how they are used.

4.1 What Is an Operator?

The allow processing of data types and objects. They are literals used to perform logical, relational or mathematical operations on the operands.

Operators in Python can be separated into following five categories:

i. Arithmetic
ii. Logical
iii. Comparison
iv. Assignment
v. Membership operators
vi. Identity operators

4.2 Arithmetic Operators

Arithmetic operators are used to perform simple mathematical operations on the operands.

Here is a list of arithmetic operators in Python

Operator	Name	Function	Example
+	Addition	For adding operands	2 + 2
-	Subtraction	Subtracts the operands	5 - 1
*	Multiplication	Multiplies the operand	5 * 8
/	Division	Divides the operand of the left by the in the right	4 / 2
%	Modulus	Divides the operand on the left by the on the right and returns a remainder	5 % 2
**	Exponent	Takes exponent of the operand on the left to the power of right	5 ** 2
//	Floor Division	Divides the operand on the left by the on the right and returns the result in which the digits after the decimal point are removed.	5 // 2

Example

```
a = 5
b = 4
print (a + b)   # 9
print (a - b)   # 1
print (a * b)   # 20
print (a / b)   # 1.25
print (a % b)   # 1
print (a ** b)  # 625
print (a // b)  # 1
```

4.3 Logical Operators

Logical operators are operators that work with Boolean data types and Boolean expressions.

Here are the logical operators in Python.

Operator	Functionality	Example
and	If both statements are true, it returns True	x == 7 and x > 13
or	If one of two expressions are true, it returns True	x == 7 or x > 13
not	Used to reverse the logical state of its operand, returns True if result is	not(x == 7 and x > 13)

	False	

Example:

```
1  a = True
2  b = False
3  print(a and b)  # False
4  print(a or b)   # True
5  print( not b)   # True
6  print(b or True) # True
7  print((5 > 9) ^ (a == b))  # False
8
```

4.4 Comparison Operators

Comparison operators in Python are used to compare two or more operands. Python supports the following comparison operators:

Operator	Name	Example
==	Equal	(2 + 1) == 3
!=	Not equal	2 != 3
>	Greater than	3 > 2
<	Less than	2 < 2
>=	Greater than or equal	(2 + 1) >= 3
<=	Less than or equal	(2 + 1) <= 3

Example:

```
1    x = 9
2    y = 10
3
4    print('x > y  is', x > y)   # False
5    print('x < y  is', x < y)   # True
6    print('x == y is', x == y)  # False
7    print('x != y is', x != y)  # True
8    print('x >= y is', x >= y)  # False
9    print('x <= y is', x <= y)  # True
```

4.5 Assignment Operators

Assignment operators allow assigning values to variables.

Operator	Example	Same as
=	y = 5	y = 5
+=	y += 5	y = y + 5
-=	y -= 5	y = y - 5
*=	y *= 5	y = y * 5
/=	y /= 5	y = y / 5
%=	y %= 5	y = y % 5
//=	y //= 5	y = y // 5
**=	y //= 5	y = y // 5
&=	y //= 5	y = y // 5

|=	y |= 5	y = y | 5
^=	y ^= 5	y = y ^ 5
>>=	y >>= 5	y = y >> 5
<<=	y <<= 5	y = y << 5

4.6 Identity Operator

Identity operators are used to check if two values (or variables) are located on the same part of the memory. Two variables that are equal does not imply that they are identical.

Operator	Description	Example
is	Returns True if variables on both sides of the operator are the same object	y is z
is not	Returns True if variables on both sides of the operator are not the same object	y is not z

Example

```
x = 5
y = 5
x2 = 'Beans'
y2 = 'Beans'
x3 = [2,2,2]
y3 = [2,2,2]

print(x is not y)   # False
print(x2 is y2)     # True
print(x3 is y3)     # False
```

4.7 Membership Operators

Membership operators are used to test whether a value or variable is found in a sequence (string, list, tuple, set and dictionary).

Operator	Description	Example
in	Returns True if value/variable is found in the sequence	5 in z
not in	Returns True if value/variable is not found in the sequence	y not in z

Example

```
1  x = 'Membership operator'
2  y = {1:'a',2:'b'}
3
4  print('h' in x) # True
5  print('Operator' not in x) # True
6  print(1 in y) # True
7  print('a' in y) # False
8
```

4.8 Bitwise operators

Bitwise operators act on operands as if they were string of binary digits. It operates bit by bit, hence the name.

Operator	Meaning	Example
&	AND	Set each bit to 1 if both bits are 1
|	OR	Set each bit to 1 if one of bits is 1
~	NOT	Inverts all bit
^	XOR	Set each bit to 1 if only one of two bits is 1
>>	Right-Shift	Moves the bits the number of positions specified by the second operand in the right direction.
<<	Left-Shift	Moves the bits the number of positions specified by the second operand in the left direction.

Example

```
1     a = 3  # 0000 0011 = 3
2     b = 5  # 0000 0101 = 5
3
4     print(a | b)    # 0000 0111 = 7
5     print(a & b)    # 0000 0001 = 1
6     print(a ^ b)    # 0000 0110 = 6
7     print(~a & b)   # 0000 0100 = 4
8     print(a << 1)   # 0000 0110 = 6
9     print(a << 2)   # 0000 1100 = 12
10    print(a >> 1)   # 0000 0001 = 1
11
```

Exercises

1. Write an expression that checks whether an integer is odd or even.

2. Write a Boolean expression that checks whether a given integer is divisible by both 5 and 7, without a remainder.

3. Write an expression that looks for a given integer if its third digit (right to left) is 7. 4. Write an expression that checks whether the third bit in a given integer is 1 or 0.

CONDITIONAL STATEMENT

In programming, conditional statement is a feature of a programming language, which perform different actions depending on whether a specific Boolean constraint evaluates to true or false. That is to say, conditional statements are used to control the flow of a program.

In Python, there are three main types of conditional statements

- If
- else
- elif

5.1 The "if" statement

Decision making is required when we want to execute a code only if a certain condition is satisfied. In programming, the if statement is used for decision making. In "if" statement, the body of the code is executed only when the if condition evaluates to True.

The syntax of the if statement is;

if test expression:
    statement(s)

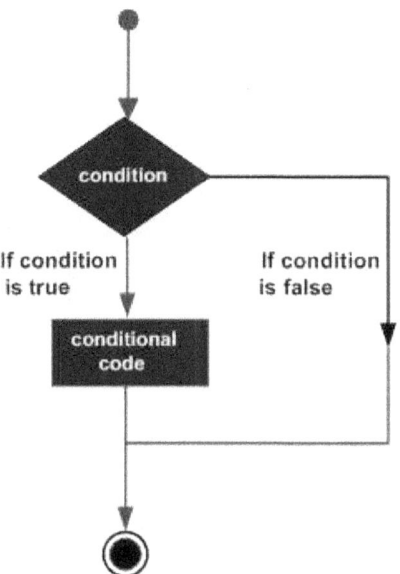

Let's see an example:

```
>>> x = 5
>>> y = 4
>>> if (x>y):
        print("Print Yes")
```

Output:

Print Yes

In the above example, x > y is the test expression. The body of the "if" is only executed only if this evaluates to True. In this case, the variable x is greater than y, the test expression is true and the body of the if is executed by displaying "**Print Yes**".

Let's take a look at another example:

>>> pass_score = 50
>>> if (pass_score > 100) :
>>>   print("You did score very high")
    print("Check out this line")

Output:

Check out this line

From the example above, the second print() statement is outside the "if" block, that is, it is not indented. Hence, it will execute regardless whether the test expression is true or false.

5.2 The "If-Else" statement

The "if" block executes only if the test expression (condition) that follows it returns true. The "else" statement is used if we want to execute alternate set of statements if the conditional expression in the if statement returns false. The else statement is an optional statement. However, there should be at most on **else** statement following **if**.

The syntax of the if..else statement is;

if (test expression):
   statement(s)
else:
   statement(s)

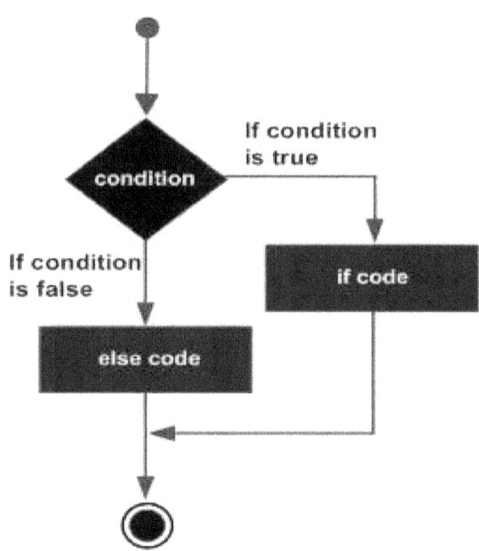

Take a look at the following example:

You want to print either pass or fail depending on a student's test score.

```
>>> score = 50
>>> if (score > 50):
        print("Passed")
else:
        print("Fail")
```

Output:
```
Fail
```

From the example, we can see that the test expression "score > 50" when evaluated returns false. Thus, the statement in the "if" block is ignored, while the "else" block is executed.

5.3 The "If-Elif-Else" Statement

In the elif statement, a user can decide among multiple selections. Here the statements are executed from top to bottom. As soon as one of the test expressions return TRUE, the statement with that if block is executed, and the rest skipped. If it happens that none of the test expression or condition returns TRUE, the final else statement will be executed.

The syntax of the elif statement is;

if (test expression 1):
   statement(s)
elif (test expression 2):
   statement(s)
.
.
else:
   statement(s)

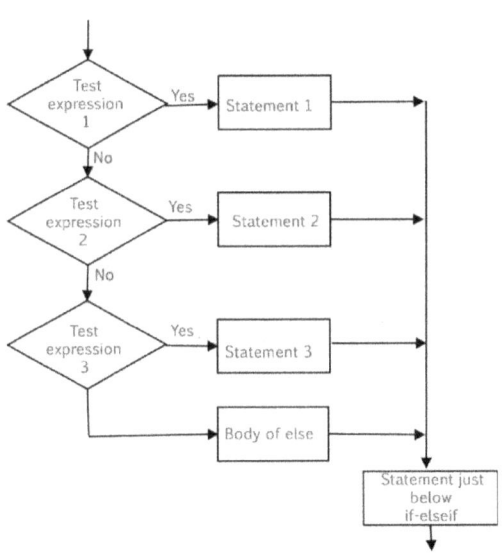

Example:

```
>>> score = 60
>>> if (score < 60):
        print("Grade is Fail")
elif (score == 60):
        print("Grade is Average")
elif ( score > 60):
        print("Grade is Excellent")
else:
        print("Grade is not present")
```

Output:

```
Grade is Average
```

Exercises

1. When checking multiple conditions in Python, which of the statements would you use and why?
2. A school has following rules for grading system:
   Below 45 - F
   45 to 49 - D
   50 to 59 - C
   60 to 69 - B
   Above 69 - A
   Ask user input and print the corresponding grade.
3. Write a program to check whether a number is even or odd.
4. Write a program to check whether an alphabet is vowel or consonant using.

LOOPS

A loop is a programming construct that allows repeated execution of a fragment of source code. Loops basically allows the execution of a statement or a group of statements multiple time. In order to enter a loop, there are certain conditions that are defined at the beginning. When the condition becomes false, the loop stops and the control moves out of the loop.

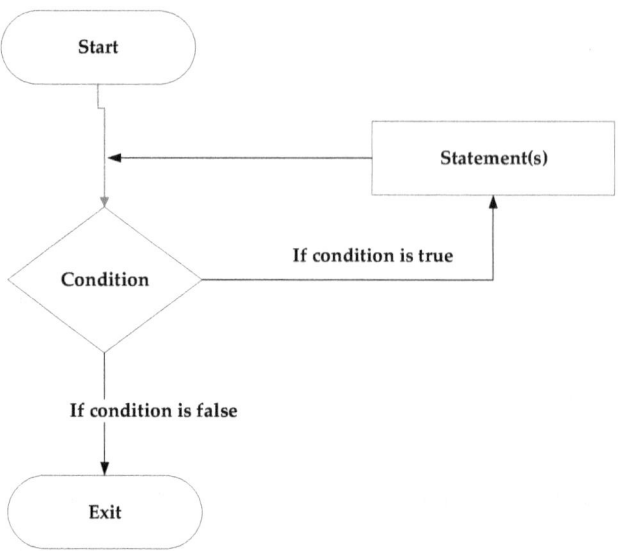

From the flowchart, the control enters from start and checks the condition, so if this condition is true, it goes on to execute the statement(s). After executing, it goes back again to check the condition, if the condition is still true, it executes the statement(s) present inside the loop. This process keeps on repeating until the condition returns false. The moment the condition returns false, the control will move out of the loop and execute the statement(s) that are present after the loop.

It should be noted that there are two kinds of loop; finite loop, which is represented on the flowchart above and infinite loop. In infinite loop, the condition will never return false, so the control will not come out of the loop, that is to say the loop never stops.

Another way of categorizing loops is; posttest and pretest. In posttest loops, the control will first enter the loop and then check the condition at the end. But in pretest loop, the control enters the loop only when the condition is true.

In posttest loop the condition is checked at the end of the loop, while in pretest loop the condition is checked at the beginning of the loop. Python only has pretest loops.

6.1 Types of loops in Python

Python supports 3 kinds of loops;

i. While
ii. For
iii. Nested

6.1.1 While Loop

While loops are known as indefinite or conditional loops. It is basically used when the number of iterations required is not known. The control keeps iterating until certain conditions are met.

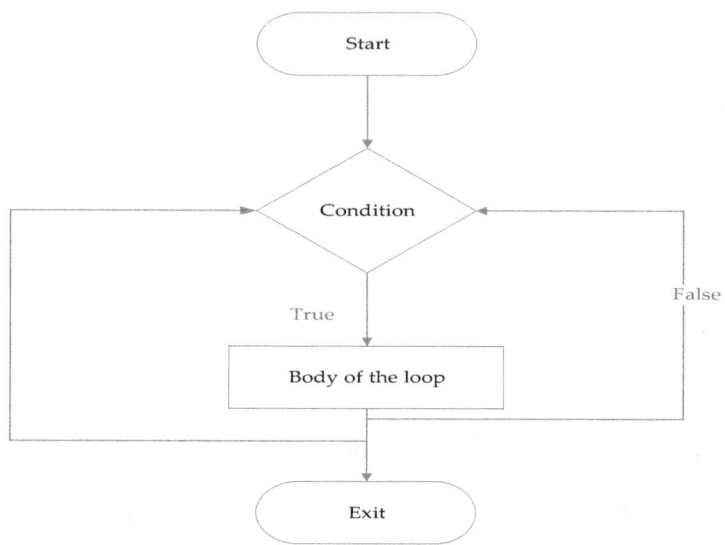

With the help of the above flowchart, the control enters the loop only when the **while condition** is true, and when it is true the body of the loop is executed. Then it goes back to check if the condition is still true or false, if the condition still returns true, it will keep on executing the body of the loop until the condition is false. The moment the condition becomes false, the control will move out of the loop and execute the statements after the loop.

The syntax of while loop is;

```
while expression:
    statement(s)
```

Example using while loop;

1. Let is consider printing the integer values between 1-10.

```
x = 1
while x < 10 :
    print ("Number: ", x)
    x = x + 1
print ("Well-done")
```

Output:

```
Number:  1
Number:  2
Number:  3
Number:  4
Number:  5
Number:  6
Number:  7
Number:  8
Number:  9
Well-done
```

2. In this example we will examine how by using the while loop we can find the sum of the numbers from 1 to n.

```
n = int(input("Enter maximal number :"))
num = 1
sum = 1
print("The sum 1", end =" ")   # end =" " is a format to prevent newline
while (num < n):
    num = num + 1
    sum += num
    print(" + " + str(num), end =" ")
print(" = " + str(sum))
```

Output:
```
Enter maximal number : 8
The sum 1  + 2  + 3  + 4  + 5  + 6  + 7  + 8  = 36
```

3. This is a popular game kids play, it called guess the correct number. A random number is chosen, when the inputted number is less than the chosen random number, print **Number too small**, if the number is larger than the random number, print **Number too Large.** When the correct number is entered, print **Yes number is correct**.

```python
import random
n = 25 #generate number range between 0-26
to_be_guessed = int(n + random.random()) + 1  #int to generates the random number
guess = 0
while guess != to_be_guessed:
    guess = int(input("Guess the number:"))
    if guess > 0:
        if (guess > to_be_guessed):
            print("Number too large")
        elif (guess < to_be_guessed):
            print("Number too small")
    else:        #come out of the loop
        print("Sorry, you are not good at this game")
        break
else:   #when answer is correct
    print("Yes number is correct")
```

We can see that from this example, the number of iterations is not known.

4. Example to check if a number is prime.

```
import math
num = int(input("Enter a positive number: "))
divider = 2 # the smallest possible divider
max_divider = math.sqrt(num) # the maximum possible divisor
prime = True # Boolean variable
# While passing through the loop, if it turns out that the
# number has a divisor, the value of prime will become false.
while (prime & (divider <= max_divider)):
    if (num % divider == 0):
        prime = False
    divider = divider + 1
print("Prime? " + str(prime))
```

Output:
```
Enter a positive number: 100
Prime? False
```

6.1.2 For Loop

In Python, for loop is used when you know the number of iterations that is required. The for loop syntax provides 3 basic information:

- Boolean condition
- The initial value of the counting variable
- Incrementation of counting variable

The syntax of while loop is;

```
For <variable> in <range>:
    statement 1
    statement 2
    ...
    statement n
```

From the flowchart below, the control enter the program, it sees items from the sequence, it will execute the statement(s), go back again and then from the range it will pick up the next item, it will keep on doing this until there are not items left in the sequence, then the statement(s) after the for loop will be executed.

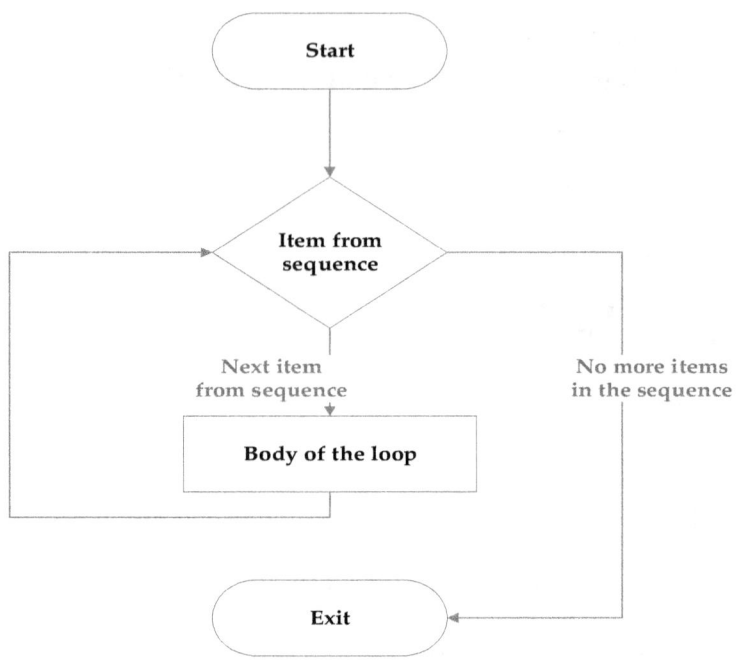

Example:

cars = ['BMW', 'Volvo', 'Mercedes', 'Toyota'] # *List of car brands*
for car in cars: # *for loop 4 iterations*
   print(car)
print ("That is for loop") # *moves out of loop*

Output:

```
BMW
Volvo
Mercedes
Toyota
That is for loop
```

2. Now we are going to look at factorial calculations. Since in factorial we already know the number of times we are going to multiply, for 4! We are going to multiply 4 times, that's 4x3x2x1 which requires only three iterations.

```
num = int(input("Enter the number:")) #input number
factorial = 1 #initial number
if (num < 0): # if condition for positive number
   print("Only positive integers")
elif (num == 0):
   print("Factorial for 0 = 1")
else:
   for i in range(1,num + 1):
      factorial = factorial * i
print(factorial)
```

6.1.3 Nested Loops

Python allows the use of a loop inside another loop. This is called nested loop. The syntax of nested loop is;

```
for variable in sequence:
    for variable in sequence:
        statement(s)
    statement(s)

while expression:
```

```
while expression:
    statement(s)
statement(s)
```

You can use a while loop inside a while loop and a for loop inside a for loop. At the same time, you can use a while loop inside a for loop or a for loop inside a while loop.

Example:

1. A program to print Pythagorean numbers. $a^2 + b^2 = c^2$

from math **import** sqrt # *import square root function*
m = int(input("Maximal Number: ")) # *print pythagorean numbers from 1 to m*
for a in range(1,m+1): # *for loop 1 to m+1, but does not include m+1*
    for b in range(a,m): # *for loop a to m-1*
        c_square = a**2 + b**2 # *c square*
        c = int(sqrt(c_square)) # *square root of c square*
        if ((c_square - c**2) == 0): # *check if c_square minus c to power 2 is zero*
            print(a,b,c)

Output:
```
Maximal Number: 10
3 4 5
6 8 10
```

2. Let's say you want to simulate the working of room reservation at a guest house.

```python
reservation = input("Make reservation, yes or no:")
while reservation == 'yes':
   num = int(input("Enter number of people:"))
   for num in range(1,num+1):
      name = input("Enter name: ")
      age = input("Enter age: ")
      room_type = input("Enter room type: ")
      print(name)
      print(age)
      print(room_type)
   reservation = input("Forgot a reservation? :")
```

Output:
```
Make reservation, yes or no:yes
Enter number of people:2
Enter name: Oladele
Enter age: 12
Enter room type: family
Oladele
12
family
Enter name: Maria
Enter age: 23
Enter room type: Delux
Maria
23
Delux
Forgot a reservation? :no
```

Exercises

1. Write a program that prints the numbers from 1 to N.

2. Write a program that reads from the console a series of integers and prints the smallest and largest of them.

3. Write a program that reads from the console number N and print the sum of the first N members of the Fibonacci sequence: 0, 1, 1, 2, 3, 5, 8, 13, 21, 34, 55, …

4. Write a program that by a given integer N prints the numbers from 1 to N in random order using while loop.

ARRAYS

An array is basically a data structure which can hold more than one value at a time. It is a collection or ordered series of elements of the same type.

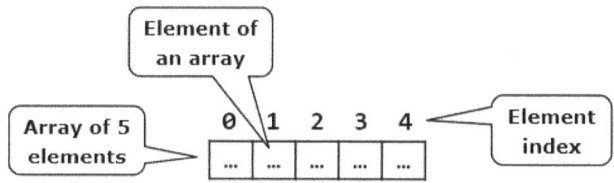

Array element index in python are numbered with 0, 1, 2, … N-1. The total number of elements in a given array we call length of an array.

7.1 Creating Arrays

Arrays in python can be created after importing the array module. The array module can be imported using three ways:

- Without alias: **import array**
- Using alias: **import array as arr**
- Using *: **from array import ***. What this does, is that it imports all that is present in the array module.

```
1  import array as arr
2  a = arr.array('i',[10,12,15,16])
```

The sample script above creates an array of integer numbers. The letter 'd' is a type code. This determines the type of the array during creation.

Commonly used type codes:

Code	Python Type
'b'	int
'B'	int
'h'	int
'H'	int
'i'	int
'I'	int
'l'	int
'L'	int
'f'	float
'd'	float

7.2 Access to the Elements of an Array

We can access the array elements directly using their index values. Each element can be accessed through the element's index (consecutive number) placed in the brackets.

Indexing starts at 0 and not from 1. Thus, the index number is always 1 less than the length of the array. Negative index values can be used as well. It is of important note to remember that negative indexing starts from the reverse, that is from right to left.

```
import array as arr
a = arr.array('i',[10,12,15,16])

print("First element:", a[0])
print("Second element:", a[1])
print("Last element:", a[-1]) # using negative index
```

7.3 Basic Array Operation

Arrays are mutable, which means they are changeable. You can easily and or remove elements from an array.

7.3.1 Length of An Array

The length of an array is the number of elements that are present in the array. To find the length of an array, we make use of len() function. The len() function returns an integer value that is equal the number of elements in the array.

The syntax is *len(array_name)*

Example:

```
import array as arr
a = arr.array('d',[10.01,12.23,15.34,16.56])

print(len(a))  # Output: 4
```

7.3.2 Add Elements to Array

To add elements to an array, you can do so using the append(), extend() and insert() function.

i. The append() function is used when you want to add a single element at the end of the array.
ii. We use the extend() function when we want to add more than one element at the end of an array.

iii. To add an element at a specific position in an array, we make use of the insert() function.

Example showing the implementation of append(), extend() and insert() functions:

```
import array as arr
a = arr.array('d',[10.01,12.23,15.34,16.56])
a.append(4.5)
print("Array a =" ,a)

b = arr.array('i',[1,2,3,4])
b.extend([5,6,7])
print("Array b" ,b)

c = arr.array('d',[10.01,12.23,15.34,16.56])
c.insert(2,13.10)    # inserts into index 2
print("Array c" ,c)
```

Output: The output is shown below

Array a = array('d', [10.01, 12.23, 15.34, 16.56, 4.5])
Array b array('i', [1, 2, 3, 4, 5, 6, 7])
Array c array('d', [10.01, 12.23, 13.1, 15.34, 16.56])

7.3.3 Remove Elements from Array

The pop() or remove() functions are used to remove elements of an array.

The pop() function is used when you want to remove an element and return it. Whereas the remove() function is used when you want to remove a specific value without returning it.

The pop() function can either take one parameter or no parameter. The parameter it takes is the index value of the element to be removed, if you do not specify any parameter, it will remove the last element from the array.

The remove() function takes one parameter, which is the element to be removed.

Example code to show how you can remove elements using these functions:

```
import array as arr
a = arr.array('d',[10.01,12.23,15.34,16.56])

print("Popping last element a =" ,a.pop())
print("Popping 2nd element a =" ,a.pop(2))

a.remove(10.01)
print(a)
```

Output: The output is shown below

```
Popping last element a = 16.56
Popping 2nd element a = 15.34
array('d', [12.23])
```

7.3.4 Array Concatenation

Array concatenation can be done using the + symbol.

```
import array as arr
a = arr.array('d',[10.01,12.23,15.34,16.56])
b = arr.array('d',[1.01,12.1,12.35])
c = arr.array('d')
c = a+b
print("Array c = ",c)
```

Output
```
Array c =  array('d', [10.01, 12.23, 15.34, 16.56, 1.01, 12.1, 12.35])
```

It should be noted that you cannot concatenate arrays of different type.

7.3.5 Slicing Array

Array slicing means fetching some particular values from your array. An array can be sliced using the : symbol. This returns a range of elements that we have specified by the index numbers.

```
import array as arr
a = arr.array('d',[10.01,12.23,15.34,16.56,20.20])
print(a[0:3])
```

From the sample script above, we created a sample array a, and slice it from zero to three [0:2]. Zero specifies where fetching has to start, while three specifies where fetching has to stop.

```
array('d', [10.01, 12.23, 15.34])
```

From the, we can see it starts from 0 and then go on till 3 without including the value which is present at 3. From the output above, we have a[0], a[1], a[2].

7.3.6 Iterating through an Array

As we can see, the iteration through the elements of an array is one of the most used techniques when we work with arrays. Consecutive iterating using a loop will allow us to access each element through its index and we will be able to modify it as we want. We can loop through an array easily using the for and while loops.

- The for loop iterates over the items of an array specified number of times.
- The while loop iterates over the elements until a certain condition is true.

When using the while loop, you have to keep three things in mind:

i. Initializing the iterator
ii. Specify a condition
iii. Increment the iterator

Example: In the following example we will double the values of all elements of an array of numbers and we will print them

```
import array as arr
a = arr.array('i',[1,2,3,4,5])

print("Output: ", end="")
for x in a:
    x = 2 * x
    print (x, end=" ")
```

What the loop does is that it goes through every element in array a and multiplies them by 2.

Output
```
Output: 2 4 6 8 10
```

For example, we can iterate through some of the elements of the array, not through all of them.

```
import array as arr
a = arr.array('i',[1,2,3,4,5])
b = 0

while b<len(a):
    print(a[b]*a[b])
    b = b+2
```

Output

1 9 25

FUNCTIONS

A function is a block of organized, reusable code that is used to perform some task. It is usually called by its name when its task need execution. Values can be passed to it, or you can have it return results.

Functions reduce lines of code in your main program by letting you avail predefined features multiple times without having to repeat its set of codes again.

8.1 Importance of Function

There are many reasons one should make use functions. Some of them are listed below.

1. Better Structured Program and More Readable Code

Whenever a program has been created, it is a good practice to use functions, in a way to make your code better structured and easy to read, so that it can be maintained by other people.

2. Avoid Duplicated Code

Another reason to use functions is that functions help us to avoid repeating code.

3. Code Reuse

If a block of code is used more than once in a program, it is a good practice to separate it in a function, which can be called many times – thus enabling reuse of the same code, without rewriting it.

8.2 Function Declaration

Def keyword is used before the name of a function, followed by parentheses before a colon (:).

Function syntax

```
def function_name(parameters):
    """docstring"""
    statement(s)
```

Docstring: This is the first string after the function header, they to describe what the function does.

Let's write a simple function that prints 'Welcome to Python" on screen.

```
>>> def print_logo():
        print("Python")
        print("www.python.org")

>>> print_logo()
```

The above script creates a function named *print_logo.*

print_logo() is use to call a function. When the script executes, the print_logo function executes an output that looks like this:

```
Python
www.python.org
```

8.3 Parameters in Functions

Sometimes to solve certain problem, the function may need additional information, which depends on what the function executes.

For instance, if there is a function that has to find the area of a square, in its body there must be the block of code that finds that area (equation s = a²). Since the area depends on the square side length, to calculate that equation for each square, the function will need to pass a value for the square side length. That is why we have to pass somehow that value, and for this purpose we use parameters.

When a parameterized function is declared, our purpose is that every time we call the function, its result changes according to its input. That is to say, when a function has parameters, its behavior depends upon parameters values.

Example: Method to Show whether a Number is Positive

The function gets as input a number and according to it prints "Positive", "Negative" or "Zero":

```
number = int(input("Enter a number: "))
def PrintSign(number):
    if (number > 0):
        print("Positive")
    elif (number < 0):
        print("Negative")
    else:
        print("Zero")

PrintSign(number)
```

8.3.1 Functions with Multiple Parameters

So far, we have looked at examples for functions with parameter lists that consist of a single parameter. When a function is declared, however, it can have as multiple parameters as the function needs.

Example: If we are asked to find the maximum of two values, the function needs two parameters:

```
number1 = int(input("Enter a first number: "))
number2 = int(input("Enter a second number: "))
def PrintMax(number1, number2):
    max = number1
    if (number2 > max):
        max = number2
    print("Maximal number: " + str(max))

PrintMax(number1, number2)
```

8.4 Returning a Result from a Function

In previous examples, the function prints on the console. Functions, however, usually do not just execute a simple code sequence, but in addition they often return results.

To make a function return value, the **return** keyword must be placed in the body of the function, followed by an expression that will be returned as a result by the function.

Take a look at the following example.

```
number1 = int(input("Enter a first number: "))
number2 = int(input("Enter a second number: "))
def Multiply(number1, number2):
    result = number1 * number2
    return result
result = Multiply(number1, number2)
print(str(number1) + "*" + str(number2) + " is " + str(result
```

In the above script, we create a function **Multiply**. The function accepts two arguments, multiplies them and returns the resultant multiplication.

Output:
```
Enter a first number: 5
Enter a second number: 5
5*5 is 25
```

Let's look at a more complex example.

We are given a task to write a program that for a body temperature entered by a patient, measured in Fahrenheit degrees, the program should convert the temperature and print it in degrees Celsius, with the following message:

a. "Your body temperature in Celsius degrees is X"

b. If the measured temperature in Celsius is higher than 37°, the program should warn the user that they are ill.

The Celsius to Fahrenheit formula is: °C = (°F - 32) * 5 / 9, where respectively with °C we mark the temperature measured in Celsius, and with °F – the temperature in Fahrenheit.

```python
def ConvertFahrenheitToCelcius(temperatureF):
    temperatureC = (temperatureF - 32) * 5 / 9
    return temperatureC

temperature = float(input("Enter your body temperature in Fahrenhiet:"))
temperature = ConvertFahrenheitToCelcius(temperature)

print("Your body temperature in Celsius degree is " + str(temperature))
if (temperature >= 37):
    print("You are ill!")
```

Output:

```
Enter your body temperature in Fahrenhiet:90
Your body temperature in Celsius degree is 32.22222222222222
```

8.5 Default Arguments

There is a way of representing syntax and default values for function parameters in Python. Default values means that the function parameter will take a value if no parameter value is passed during function call. Take a look at the following script:

```python
def staff(firstname, lastname ='Chow', department ='Sales'):
    print(firstname, lastname, 'works in', department, 'Department')

staff(firstname ='Kingsley')
staff(firstname ='John', department ='Human Relations')
staff(lastname ='Tony', firstname ='Blair')
```

Output:

```
Kingsley Chow works in Sales Department
John Chow works in Human Relations Department
Blair Tony works in Sales Department
```

In the first call, there is only one required keyword argument. In the second call, one is required argument and one is optional(department), whose value get replaced from default to new passing value. We can see in the third call, that order in keyword argument is not important.

8.6 Anonymous Functions

Anonymous functions are functions that do have any name, they are also known as Lambda or nameless functions. The keyword 'lambda' is used for creating anonymous functions.

The main purpose of lambda functions comes in to play when you need some functions just once, they are created when needed. Due to this reason, python lambda functions are known as throw-away functions. They are also used within higher-order functions which takes a function as input or return it as output.

8.6.1 How to write Anonymous Functions

A lambda function is created using the lambda keyword. The syntax is as follows:

lambda arguments: expression

Example: To find square of a number

```
>>> x= lambda a: a*a
>>> x(3)
```

8.6.2 Lambda within user-defined functions

Lambda functions are bet used with other higher-order functions.

Example

```
1   def Test(a):
2       return(lambda b:a+b)
3   t = Test(4)
4   print(t(8))
5
```

From the example script above, the lambda function that is used within the function Test is called whenever the higher-order function is called. The expression *t = Test(4)* for passing a value to the variable **a.** print(t(8)) prints the value of a + b.

8.7 Scope of Variables

There are two scope of variables in Python; global and local. Scope of a variable refers to the part of code where a variable can be assessed. Global variables are variables that are declared outside the function and can be used anywhere in the program. While local variables are variables declared inside a function and it cannot be accessed outside the function.

Take a look at the example below:

```
x = 50  # global variable

def Print_Number():
    y = 500    # Local variable
    z = y - x  # global variable used
    print(y)
    print(z)

print(x)
Print_Number()
```

As you can see from the script x is present everywhere in the code and was even used inside the function. If we were to use y outside the function, it will throw an error because y is

73

a local variable. Outside the function, we can also access x.

The output is shown below

```
50
500
450
```

FILE HANDLING

In Python, users do not need to import an external library to read and write files. Python provides an inbuilt file handling function that allows users to handle files that is; to create, read and write files.

In this chapter, we will learn

- How to Create a Text File
- How to Append Data to a File
- How to Read a File
- How to Read a File line by line
- File Modes in Python

In Python, a file operation takes place in the following order.

- Open a file
- Read or write (perform operation)
- Close the file

9.1 How to open a file?

To open a file for writing or use in Python, you must make use of the built-in open() function. This function returns a file object, also referred to as a handle, because it is used to read or modify the file. The open() takes two parameters, filename and mode. The syntax is *open(filename, mode)*

The filename is the file that you want to open. While the mode specifies the type of operation to be perform. Below is the list of mode:

'r'	**Read** – Default value. Open a file for reading. Error if the file does not exist.
'w'	**Write** - Open a file for writing. Creates a new file if it does not exist or truncates the file if it exists.
'x'	**Create** – Creates the specified file, if the file already exists, the operation fails.
'a'	**Append** – Opens file for appending at the end of the file without truncating it. Creates a new file if it does not exist.
't'	Default value. Open in text mode.
'b'	Open in binary mode. For example, images
'+'	Open a file for updating (reading and writing)

Example

```
f = open("list.txt")
f = open("list.txt",'r')
```

The above two code snippets are the same.

9.2 Reading File

There are lot of ways to read a text file in python. You can choose to read all the characters or just some characters contained in the text file.

Example

In this example we have a txt file name list on our desktop with contents as 0123456789

```
import os
file = open("C:/Users/Isujeh/Desktop/list.txt", 'r')
print(file.read())
file.close()
```

The above code reads everything that is available in the list.txt file

Output

0123456789

Let's that a look at another example while still making use of the list.txt file.

```
1  import os
2  file = open("C:/Users/Isujeh/Desktop/list.txt",'r')
3  print(file.read(4))
4  file.close()
```

From this example *file.read(4)* reads only the first four characters The output is 0123.

In the previous examples, we have learnt how to read characters. Now we are going to learn how to read lines.

```
1  import os
2  file = open("C:/Users/Isujeh/Desktop/list2.txt",'r')
3  print (file.readline()) #Line by line output
4  file.close()
5
6  file = open("C:/Users/Isujeh/Desktop/list2.txt",'r')
7  print(file.readlines()) #Read lines separately
8  file.close()
```

Output

```
Tom Marvolo Riddle

['Tom Marvolo Riddle\n', 'I am Lord Voldemort\n', 'Harry Potter']
```

9.2 File Write Method

To write to an existing file, you must add a parameter to the open() function:

"a" – Append – will appended to the end of the file.

"w" – Write – will overwrite any existing content

Example

```
1  import os
2  file = open("C:/Users/Isujeh/Desktop/list2.txt",'a')
3  file.write("I love learning python!!!")
4  file.close()
```

```
1  import os
2  file = open("C:/Users/Isujeh/Desktop/list2.txt",'w')
3  file.write("Ooops overwritten !!!")
4  file.close()
```

9.3 Creating a New File

To create a new file in python, the open() function, with one of the following parameters is used:

"x" – Create – will create a file, returns an error if the file exists.

"a" - Append – will create a file if the specified file does not exist.

"**w**" – Write – will create a file if the specified file does not exist.

```python
import os
file = open("C:/Users/Isujeh/Desktop/new.txt",'w')
file.write("New file")
file.close()
```

9.4 Delete a File

To delete a file, you must import the OS module, and run its os.remove() method:

```python
import os

if os.path.exists("C:/Users/Isujeh/Desktop/new.txt"):
    os.remove("C:/Users/Isujeh/Desktop/new.txt")
else:
    print("The file does not exist!")
```

OBJECT-ORIENTED PROGRAMMING (OOP)

Object-oriented programming is the successor of procedural programming. The object-oriented approach relies on the paradigm that each and every program works with data that describes entities (objects or events) from real life. For example: consider a scenario that you were asked to design a car, the car would have a name, color, wheels, size, length, etc. This is how objects came to be. They describe the properties or characteristics and behavior (methods) of such real-life entities. OOP enables the easy reuse of code by applying simple and widely accepted principles.

10.1 Fundamental Principles of OOP

For a programming language to be object-oriented, it has to enable working with classes and objects as well as the implementation and use of the fundamental object-oriented principles and concepts: principles of OOP:

- Encapsulation We will learn to hide unnecessary details in our classes and provide a clear and simple interface for working with them.

- Inheritance We will explain how class hierarchies improve code readability and enable the reuse of functionality.

- Abstraction We will learn how to work through abstractions: to deal with objects considering their important characteristics and ignore all other details.

- Polymorphism We will explain how to work in the same manner with different objects, which define a specific implementation of some abstract behavior.

10.2 Classes

A class is the blueprint from which objects are created. Classes are a description (model) of real objects and events referred to as entities. Objects are instances of classes. For example, Maurice is a Student and Alex is also a Student.

The syntax for how to write a class

```
class ClassName:
    <statement-1>
    .
    .
    .
    <statement-N>
```

The three major aspect to a class are;

Class Variable: A variable that is shared by all instances of a class.

Instance Variable: Instance variable unique to each instance.

Data Member: A class variable or instance that holds data associated with a class and its objects.

10.2.1 Attributes and Methods in class

A class by itself is of no use unless there is some functionality associated with it. Functionalities are defined by setting attributes, which act as containers for data and functions related to those attributes. Those functions are called methods.

Attributes

You can define the following class with the name Car. This class will have an attribute name.

```
>>> class Car:
        name = "Toyota" # set an attribute `name` of the class
```

You can assign the class to a variable. This is called object instantiation. You will then be able to access the attributes that are present inside the class using the dot(.) operator. For example, in the Car example, you can access the attribute name of the class Car.

```
>>> # instantiate the class Car and assign it to variable car
>>> car = Car()
>>>
>>> # access the class attribute name inside the class Car
>>> print(car.name)
Toyota
```

Methods

Once there are attributes that "belong" to the class, you can define functions that will access the class attribute. These functions are called methods. When you define methods, you will need to always provide the first argument to the method with a **self** keyword.

For example, you can define a class `Car`, which has one attribute `name` and one method `change_name`. The method change name will take in an argument `new_name` along with the keyword `self`.

```
class Car:
    name = "Toyota"
    def change_name(self, new_name):
        self.name = new_name    # access the class attribute
with the self keyword

# Now, you can instantiate this class Snake with a variable
snake and then change the name with the method change_name.

# instantiate the class
car = Car()
print(car.name)

# change the name using the change_name method
car.change_name("BMW")
print(car.name)
```

Output:

Toyota
BMW

PYTHON CONSTRUCTOR

A constructor is a special type of method (function) which is used to initialize the instance members of the class.

Constructors can be of two types.

- Parameterized Constructor
- Non-parameterized Constructor

Constructor definition is executed when we create the object of this class. Constructors also verify that there are enough resources for the object to perform any start-up task.

11.1 Creating the constructor in python

In python, the method **__init__** simulates the constructor of the class. This method is called when the class is instantiated. We can pass any number of arguments at the time of creating the class object, depending upon __init__ definition. It is mostly used to initialize the class attributes. Every class must have a constructor, even if it simply relies on the default constructor.

Consider the following example to initialize the Student class attributes.

```python
class Student:
    def __init__(self,name,id):
        self.id = id;
        self.name = name;

    def display (self):
        print("ID: %d \nName: %s"%(self.id,self.name))

stud1 = Student("John Mike",101)
stud2 = Student("David Lee",102)

# accessing display() method to print student 1 information
stud1.display();

# accessing display() method to print student 2 information
stud2.display();
```

Output:

```
ID: 101
Name: John Mike
ID: 102
Name: David Lee
```

Non-Parameterized Constructor Example

```
class Student:
    # Constructor - non parameterized
    def __init__(self):
        print("This is non parametrized constructor")

    def show(self,name):
        print("Hello",name)

student = Student()
student.show("John Snow")
```

Output:

```
This is non parametrized constructor
Hello John Snow
```

Parameterized Constructor Example

```
class Student:
    # Constructor parameterized
    def __init__(self,name):
        print("This is parametrized constructor")
        self.name = name

    def show(self):
        print("Hello",self.name)

student = Student("John Snow")
student.show()
```

Output:

This is parametrized constructor
Hello John Snow

11.2 Python In-built class functions

The in-built functions defined in the class are described in the following table.

SN	Function	Description
1	getattr(obj,name,default)	It is used to access the attribute of the object.
2	setattr(obj, name,value)	It is used to set a particular value to the specific attribute of an object.
3	delattr(obj, name)	It is used to delete a specific attribute.
4	hasattr(obj, name)	It returns true if the object contains some specific attribute.

Example

1. class Student:
2.   def __init__(self,name,id,age):
3.   self.name = name;
4.   self.id = id;
5.   self.age = age
6.
7. #creates the object of the class Student
8. s = Student("John",101,24)
9.

10. #prints the attribute name of the object s
11. print(getattr(s,'name'))
12.
13. # reset the value of attribute age to 25
14. setattr(s,"age",25)
15.
16. # prints the modified value of age
17. print(getattr(s,'age'))
18.
19. # prints true if the student contains the attribute with name id
20.
21. print(hasattr(s,'id'))
22. # deletes the attribute age
23. delattr(s,'age')
24.
25. # this will give an error since the attribute age has been deleted
26. print(s.age)

Output:
John
25
True
AttributeError: 'Student' object has no attribute 'age'

11.3 Built-in class attributes

Along with the other attributes, a python class also contains some built-in class attributes which provide information about the class.

The built-in class attributes are given in the below table.

SN	Attribute	Description
1	__dict__	It provides the dictionary containing the information about the class namespace.
2	__doc__	It contains a string which has the class documentation
3	__name__	It is used to access the class name.
4	__module__	It is used to access the module in which, this class is defined.
5	__bases__	It contains a tuple including all base classes.

Example

1. class Student:
2.   def __init__(self,name,id,age):
3.   self.name = name;
4.   self.id = id;
5.   self.age = age
6.   def display_details(self):
7.   print("Name:%s, ID:%d, age:%d"%(self.name,self.id))
8. s = Student("John",101,22)
9. print(s.__doc__)
10. print(s.__dict__)
11. print(s.__module__)

Output:

None
{'name': 'John', 'id': 101, 'age': 22}
__main__

INHERITANCE

Inheritance is a fundamental principle of object-oriented programming. It allows a class to "inherit" (behavior or characteristics) of another, more general class. The class which inherits the properties is called the child class or derived class. While the class from which we inherit is referred to as parent class or base class.

```python
class Parent:
    def func1(self):
        print('This is a parent function')

class Child(Parent):
    def func2(self):
        print('This is a child function')

ob = Child()
ob.func1()
ob.func2()
```

In the example script above we create a Parent class and a Child class. The Child class inherits the Parent class. To inherit a class, you just have to pass the parent class name inside the parenthesis that follow the child class name. In the above script Parent class contains a function named *func1*, while the child class also contains a method named *func2*. We then create a child class object named *'ob'*. From this *'ob'* object we call the *func1*. You can see that though Child class doesn't contain *func1*, but since it is inheriting Parent class which contains *func1*, therefore the Child class object can also access this method. Finally, we call the Child class method *func2*.

Example:

```
class Mammal:
    def walk(self):
        print("Mammal walks")
#child class Human inherits the base class Mammal
class Human(Mammal):
    def talk(self):
        print("Human Talks")
d = Human()
d.talk()
d.walk()
```

Sub-classing

This is the process of calling a constructor of the base class by mentioning the base class name in the declaration of the derived class. A derived class identifies its base class by sub-classing.

Constructors with Inheritance

Every time a class is being used to make an object the __init__() function is called. When the __init__() function is added in a base(parent) class, the child(derived) class will no longer be able to inherit the parent class's __init__() function. The child's class __init__() function overrides the parent class's __init__() function.

```python
class Parent:
    def __init__(self, fname, fage):
        self.firstname = fname
        self.age = fage
    def view(self):
        print(self.firstname, self.age)
class Child(Parent):
    def __init__(self, fname, fage):
        Parent.__init__(self, fname, fage)
        self.lastname = "Jones"
    def view(self):
        print("Fighter", self.firstname, "is", self.age, "years old.",
              self.lastname, "has won many UFC fights")
ob = Child("John", '28')
ob.view()
```

CONCLUSION

Thanks for reading this book. I am certain that you have gained valuable knowledge in python programming that will stick for life. Don't stop here, read other books, solve questions and build on the knowledge gotten from this book.

www.ingramcontent.com/pod-product-compliance
Lightning Source LLC
Chambersburg PA
CBHW070659220526
45466CB00001B/508